Calculus Explained Like I'm 5: Calculus Made Ridiculously Simple

Roggie Clark

Continue Learning At Our Website: Roggieclarkbooks.com

Continue exploring educational resources, tools, books, workbooks, and learning materials designed to help readers continue building knowledge and practical skills. Visit: www.roggieclarkbooks.com

Inside you'll find:

- Educational worksheets
- Beginner-friendly guides
- Printable activities
- Companion materials for books and courses

A growing library of educational content for readers, collectors, learners, parents, teachers, and lifelong students.

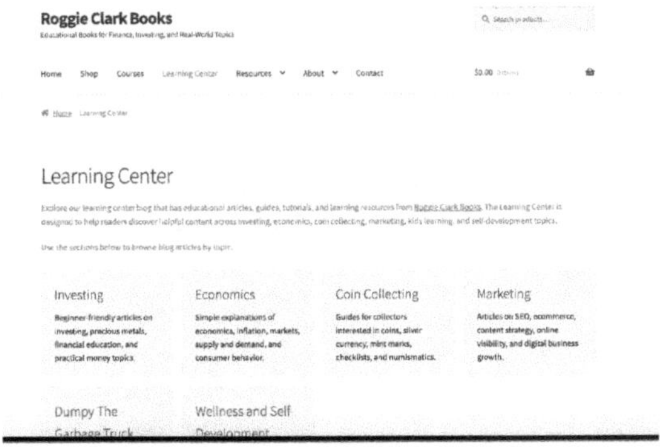

The Strategic Guide to U.S. Mint Gold and Silver Coin Investing

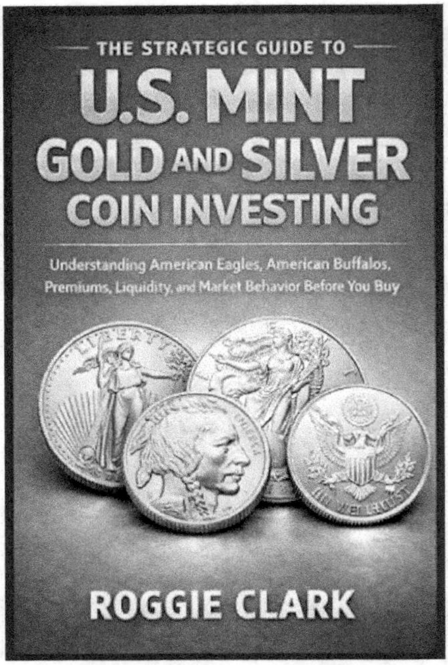

Rather than offering a general introduction to precious metals, this guide examines how American Gold Eagles, American Gold Buffalos, and Silver American Eagles behave in real investor markets.

https://www.amazon.com/Strategic-Guide-U-S-Silver-Investing/dp/B0GQKJF9K6

Copyright

Title: Calculus Explained Like I'm 5: Calculus Made Ridiculously Simple
Author: Roggie Clark

Cover Design: Roggie Clark
Editing: Renee' Clark

Printed in the United States of America

Disclaimer

The views and opinions expressed in this publication are those of the author and do not necessarily reflect the official policies, positions, opinions, or viewpoints of any individual, company, organization, institution, agency, or entity referenced or mentioned within this work.

The author is not necessarily affiliated with, endorsed by, sponsored by, or compensated by any individuals, companies, organizations, brands, institutions, products, or services referenced within this publication unless specifically stated otherwise. Any examples, sample situations, stories, illustrations, or scenarios presented may be fictionalized or simplified for educational purposes. Any resemblance to actual persons, living or dead, businesses, or events is purely coincidental unless otherwise stated.

The information, ideas, strategies, concepts, techniques, opinions, suggestions, and educational materials contained in this publication are provided on an "as is" basis without warranties of any kind, either express or implied. While every effort has been made to provide accurate and current information, neither the author nor the publisher makes any representations or warranties regarding the completeness, accuracy, reliability, suitability, or applicability of the information presented.

This publication is intended solely for educational and informational purposes and should not be interpreted as professional advice of any kind, including but not limited to financial, investment, legal, tax, business, medical, mental health, educational, psychological, or other professional guidance. Readers should conduct their own independent research and consult appropriately qualified professionals before making decisions that may affect their finances, health, business, education, legal matters, personal circumstances, or other important areas of life.

Results, outcomes, and experiences may vary significantly among individuals. Neither the author nor the publisher guarantees any specific outcomes, results, earnings, performance improvements, personal achievements, or benefits resulting from the use of information contained in this publication.

By reading, accessing, or using this publication, you acknowledge and agree that the author and publisher shall not be held liable for any direct, indirect, incidental, consequential, financial, personal, emotional, or other damages, losses, or claims arising from the use, misuse, interpretation, or application of information contained within this publication. You assume full responsibility for any actions, decisions, and outcomes resulting from the use of this material.

Image Credits and Public Domain Notice

All images, illustrations, graphics, photographs, visual references, and related materials included in this publication are believed to be in the public domain, used under applicable licenses or permissions, created by the author or publisher, or obtained from sources that clearly identify them as available for lawful use.

The author and publisher have made a good faith effort to use only visual materials that are lawfully available for use and distribution. Where applicable, images and visual materials may have been resized, cropped, edited, enhanced, stylized, modified, reformatted, or arranged for educational, informational, design, formatting, or presentation purposes.

This publication is intended for educational, informational, and illustrative purposes only. The inclusion of any image, artwork, photograph, visual reference, symbol, historical material, design, or graphic element is not intended to claim ownership over public domain materials or any third-party intellectual property rights that remain with their respective owners.

If you believe any image or visual material included in this publication has been used in error, please contact the author or publisher so the matter may be reviewed and corrected in future editions or releases if necessary.

Any factual information, specifications, historical references, values, descriptions, statistics, or related content accompanying visual materials are provided for general informational purposes only and may change over time. Readers should conduct their own independent research and consult appropriate sources or qualified professionals when making decisions based on information presented in this publication.

This publication is independently produced and is not affiliated with, endorsed by, sponsored by, or associated with any company, organization, government agency, institution, trademark holder, or third party unless specifically stated otherwise.

Table of Contents

Introduction: Calculus Is Really About Change

Calculus has a reputation for being difficult. You see symbols, strange-looking equations, and words such as limits, derivatives, and integrals, and it can feel like you are learning an entirely new language.

But the basic idea behind calculus is much simpler:

Calculus helps us understand things that change.

A car changes speed. A cup of coffee cools down. Money grows in an account. Water fills a container. A business gains or loses customers. Calculus gives us ways to describe these changes and answer questions about them.

Why Calculus Looks Harder Than It Actually Is

Calculus often feels difficult because the notation is introduced before the idea behind it is clear.

For example, imagine you drive 120 miles in two hours. Your average speed is:

120 / 2 = 60 miles per hour

Easy enough.

But what if I ask:

How fast were you traveling exactly 37 minutes into the trip?

Your average speed cannot answer that. You could have been traveling 45 mph, 60 mph, or 75 mph at that particular moment.

Calculus gives us a way to answer questions like this.

The symbols are simply a mathematical language for describing ideas such as:

- How fast something is changing
- What happens at a particular moment
- How much something has accumulated
- Where something reaches its highest or lowest value

Once you understand the question being asked, the mathematics becomes much easier to follow.

The Big Idea: Understanding Things That Change

Most introductory calculus can be understood through three major ideas.

Limits are about getting closer and closer to something.

Imagine walking toward a door. You are 10 feet away, then 5 feet, then 2 feet, then 1 foot. Your distance is getting closer and closer to zero. Limits give us a mathematical way to study this kind of approaching behavior.

Derivatives answer a different question:

How fast is something changing right now?

Your car's speedometer is a useful example. It does not tell you how far you have traveled. It tells you how quickly your position is changing at that moment.

Integrals turn the question around:

How much has accumulated?

Imagine water flowing into a bucket. If you know how quickly the water is flowing at different moments, an integral can help determine how much water accumulated over time.

So you can think of the three ideas this way:

- Limit: What are we getting closer to?
- Derivative: How fast is it changing?
- Integral: How much has accumulated?

That is not everything calculus can do, but it gives us a clear foundation.

How to Use This Book Without Getting Lost in the Math

We are going to learn real calculus, but we are not going to start by throwing formulas at you.

For each major idea, we will work in a simple order:

1. **Understand the question.** What are we trying to figure out?
2. **Understand the idea.** What is calculus doing for us?
3. **Learn the notation.** How is that idea written mathematically?
4. **Try the math.** How do we actually calculate the answer?

When you see an equation, do not immediately worry about memorizing it. First ask what it means.

For example, if you eventually see:

$f'(x)$

you will learn that this notation represents a derivative. Instead of seeing random symbols, you will be able to think:

"Okay, we are interested in how something is changing."

That is the goal of this book.

You do not need to become a mathematician. You do not need to memorize dozens of formulas at once. And you certainly do not need to pretend something makes sense when it does not.

We will build the subject one idea at a time.

Last Thoughts

Calculus is built around a surprisingly understandable idea: **change**. Limits help us understand what values are approaching, derivatives help us measure how quickly things change, and integrals help us understand accumulation. The mathematics will become more detailed as we continue, but those basic ideas will remain underneath everything we do. In the next chapter, we will cover only the math foundation we actually need so that the rest of calculus has something solid to build on.

Chapter 1: The Math You Need Before Calculus

Before learning calculus, we need a few basic math ideas. We are not going to review years of algebra or turn this into a giant refresher course. We only need enough to understand what calculus is talking about. The most important ideas are **variables**, functions, equations, graphs, and slope. Once these pieces are clear, concepts such as limits and derivatives will have somewhere to land.

Variables, Numbers, and Mathematical Relationships

A number represents a specific amount:

5

20

100

A variable is different. A **variable** is a symbol that represents a quantity that can change.

For example:

x = number of hours worked

Suppose you earn \$20 per hour. We could represent your total earnings with:

$E = 20x$

If you work 1 hour:

$E = 20(1) = 20$

If you work 5 hours:

$E = 20(5) = 100$

If you work 10 hours:

$E = 20(10) = 200$

The important idea is the relationship:

When x changes, E changes.

Calculus is full of relationships like this because calculus is interested in changing quantities.

Sometimes the relationship is simple:

$d = 60t$

If d represents distance and t represents time, this equation describes distance increasing at a constant rate of 60 miles per hour.

But relationships do not always change at a constant rate.

Consider:

$d = 5t^2$

If:

$t = 1$, then $d = 5$

$t = 2$, then $d = 20$

$t = 3$, then $d = 45$

Notice what happened.

From $t = 1$ to $t = 2$, distance increased by 15.

From $t = 2$ to $t = 3$, distance increased by 25.

The amount of change is itself changing.

That is exactly the kind of situation where calculus becomes useful.

When you see an equation, start by asking:

- What quantities are changing?
- Which quantity depends on another?
- What happens to one when the other changes?

You do not need calculus yet. You simply need to recognize the relationship.

Understanding Functions Without the Confusion

A function is easier to understand than its name makes it sound.

Think of a function as a machine:

Input -> Rule -> Output

Suppose the rule is:

"Take a number and double it."

If you put in 3, you get 6.

If you put in 5, you get 10.

Mathematically, we could write:

$f(x) = 2x$

The notation:

$f(x)$

is read as:

"f of x"

It simply represents the output of the function when the input is x.

Suppose:

$f(x) = 2x + 1$

If:

$x = 3$

then:

$f(3) = 2(3) + 1$

$f(3) = 7$

We put 3 into the function and got 7 out.

Now try:

$x = 10$

$f(10) = 2(10) + 1$

$f(10) = 21$

That is the basic idea.

Functions matter in calculus because they allow us to describe relationships between changing quantities.

For example:

s(t) = position over time

T(t) = temperature over time

C(q) = cost based on quantity

The letters themselves are not important. They simply help us remember what the quantities represent.

When you see:

f(x)

think:

"The output depends on the input."

That is enough for now.

Reading Equations and Graphs as Stories About Change

An equation describes a relationship using symbols.

A graph lets us see that relationship.

Suppose:

$y = 2x$

Some values are:

$x = 0, y = 0$

$x = 1, y = 2$

$x = 2, y = 4$

$x = 3, y = 6$

If we plot these points, they form a straight line.

Every time x increases by 1, y increases by 2.

We describe how steeply a line rises or falls using its slope.

For two points, we can calculate:

slope = change in y / change in x

For example, take:

(1,2)

and:

(3,6)

The change in y is:

$6 - 2 = 4$

The change in x is:

$3 - 1 = 2$

So:

slope = 4/2 = 2

In plain English:

y increases by 2 for every 1-unit increase in x.

Now consider:

y = x^2

Some values are:

x = 1, y = 1

x = 2, y = 4

x = 3, y = 9

This time, the graph curves.

Between x = 1 and x = 2:

slope = (4 - 1)/(2 - 1) = 3

Between x = 2 and x = 3:

slope = (9 - 4)/(3 - 2) = 5

The slope changed.

That creates an interesting question:

What is the slope at one exact point on the curve?

For a straight line, the answer is easy because the slope stays the same.

For a curve, the slope keeps changing.

We need a way to zoom in on one particular point and determine what the slope is doing there.

That question takes us directly toward calculus.

Last Thoughts

You do not need to remember every algebra rule before learning calculus. The important foundation is understanding that variables represent quantities that can change, functions describe relationships between those quantities, equations express those relationships mathematically, and graphs allow us to see how they behave. Slope is especially important because it tells us how one quantity changes compared with another. Straight lines have a constant slope, but curves can have slopes that continuously change. Figuring out what happens at one exact point on a curve requires a new idea, and that is where limits come in.

Chapter 2: Limits: Getting Really, Really Close

In the last chapter, we ended with an important question: how do we find the slope at one exact point on a curved line? Calculating the slope between two different points is easy, but a single point does not give us the usual "rise over run" calculation. Calculus solves this problem by using limits. A limit lets us look at what happens as we get closer and closer to a particular value, even when reaching that exact value creates a mathematical problem.

What It Means for a Value to Approach Something

Imagine walking toward a wall.

You start:

10 feet away

Then:

5 feet away

Then:

1 foot away

Then:

0.5 feet away

Then:

0.1 feet away

Your distance is getting closer and closer to:

0

That is the basic idea of approaching a value.

In calculus, we might ask what happens to a function as x approaches a particular number.

Suppose:

$f(x) = 2x$

What happens as x approaches 3?

Try values close to 3:

$x = 2.9$ gives $f(x) = 5.8$

$x = 2.99$ gives $f(x) = 5.98$

$x = 2.999$ gives $f(x) = 5.998$

From the other direction:

x = 3.1 gives f(x) = 6.2

x = 3.01 gives f(x) = 6.02

x = 3.001 gives f(x) = 6.002

The outputs are getting closer and closer to:

6

We write:

lim x -> 3 f(x) = 6

Read that as:

"The limit of f(x) as x approaches 3 is 6."

The key word is **approaches**.

A limit is interested in what happens as we get closer to a value. That is slightly different from asking what happens exactly at that value.

Understanding Limits With Simple Examples

Sometimes finding a limit is extremely easy.

Suppose:

f(x) = x + 5

and we want to know what happens as x approaches 2.

We can simply substitute:

f(2) = 2 + 5

f(2) = 7

So:

lim x -> 2 (x + 5) = 7

But limits become more useful when direct substitution does not work.

Consider:

f(x) = (x^2 - 4)/(x - 2)

Try putting 2 into the function:

f(2) = (4 - 4)/(2 - 2)

f(2) = 0/0

We cannot divide by zero.

But remember, the limit is not asking what happens exactly at x = 2.

It is asking:

What happens as x gets really close to 2?

We can simplify the function.

Factor:

$x^2 - 4 = (x - 2)(x + 2)$

So:

$f(x) = [(x - 2)(x + 2)]/(x - 2)$

For x values other than 2, the matching factors cancel:

$f(x) = x + 2$

Now try values close to 2.

If:

$x = 1.9$

then:

$f(x) = 3.9$

If:

$x = 1.99$

then:

$f(x) = 3.99$

If:

$x = 2.01$

then:

$f(x) = 4.01$

If:

$x = 2.001$

then:

$f(x) = 4.001$

The values are approaching:

4

Therefore:

$\lim x \to 2 \ (x^2 - 4)/(x - 2) = 4$

The original function is undefined exactly at $x = 2$, but its limit can still exist.

That gives us an important distinction:

The value of a function and the limit of a function do not always have to be the same thing.

We also care about which direction we approach from.

Imagine approaching a house on a street. You could approach it from the left or from the right.

We can do the same thing with a function.

A **one-sided limit** looks at what happens from only one direction.

If the function approaches the same value from both sides, we can have a regular two-sided limit.

For example:

From the left -> 5

From the right -> 5

Then the limit can be:

5

But suppose:

From the left -> 5

From the right -> 10

The two sides disagree.

In that situation, the two-sided limit does not exist.

So when thinking about a limit, imagine asking:

"What does the function seem to be heading toward?"

If both sides tell the same story, we have a limit. If they tell different stories, we may not.

Why Calculus Needs Limits in the First Place

Now we can return to the problem that started all of this.

Suppose:

$f(x) = x^2$

We want to find the slope exactly at:

$x = 2$

We cannot find a slope using only one point, so we start with two points.

Use:

$x = 2$

and:

$x = 3$

The corresponding points are:

$(2,4)$

and:

$(3,9)$

Calculate the slope:

slope = (9 - 4)/(3 - 2)

slope = 5

Now move the second point closer.

Use:

x = 2.1

Then:

f(2.1) = 4.41

The slope is:

(4.41 - 4)/(2.1 - 2)

= 4.1

Move even closer:

x = 2.01

The slope becomes:

4.01

Closer again:

x = 2.001

The slope becomes:

4.001

Look at the pattern:

5

4.1

4.01

4.001

The slopes are approaching:

4

So the slope of the curve at $x = 2$ is:

4

This is the big reason we needed limits.

We started with the slope between two points. Then we moved the second point closer and closer to the first.

The two points never need to become the same point during our ordinary slope calculation. Instead, we ask what the slopes approach as the distance between the points approaches zero.

That gives calculus a way to find the slope at one exact point on a curve.

And that slope has an important name:

the derivative.

Last Thoughts

A limit is simply a way of asking what a value is approaching as we get closer and closer to a particular point. Sometimes we can find a limit by direct substitution, while other problems require us to examine what is happening nearby. The most important reason limits matter for us is that they solve the problem of finding the slope at one exact point on a curve. By moving two points closer and closer together and watching what their slopes approach, we can determine an instantaneous slope. In the next chapter, we will give that idea its proper name and explore derivatives.

Chapter 3: Derivatives: How Fast Is Something Changing Right Now?

In the last chapter, we used limits to find the slope of a curve at one exact point. That slope is called a derivative. A derivative tells us **how quickly something is changing at a particular moment**. This is what makes derivatives so useful. Instead of only knowing how much something changed over a long period, we can examine its rate of change right now. Speed, growth, cooling, falling prices, rising costs, and countless other changes can all be described using derivatives.

From Average Change to Change at One Exact Moment

Imagine driving 120 miles in two hours.

Your average speed is:

120 / 2 = 60 miles per hour

But that does not mean you traveled exactly 60 mph during every second of the trip.

You might have:

- Driven 70 mph on the highway
- Slowed to 40 mph in traffic
- Stopped at a traffic light
- Accelerated back to 60 mph

Your **average rate of change** describes what happened across the entire interval.

A derivative gives us an **instantaneous rate of change**, meaning the rate at one particular moment.

Your car's speedometer provides a useful example.

If it reads:

55 mph

that does not mean you have traveled 55 miles.

It means your position is changing at a rate of 55 miles per hour at that moment.

Now suppose the position of an object is:

s(t) = t^2

where t represents time.

At:

t = 1

$s(1) = 1$

At:

$t = 2$

$s(2) = 4$

The average rate of change between those times is:

$(4 - 1)/(2 - 1) = 3$

But suppose we want the rate exactly at:

$t = 1$

As we learned with limits, we can calculate rates over smaller and smaller intervals.

From $t = 1$ to $t = 1.1$, the rate is:

2.1

From $t = 1$ to $t = 1.01$, the rate is:

2.01

From $t = 1$ to $t = 1.001$, the rate is:

2.001

These rates approach:

2

Therefore, the instantaneous rate of change at $t = 1$ is:

2

That is the derivative at that point.

Understanding Slope and the Derivative

A derivative can also be understood as the slope of a curve at a particular point.

Consider:

$f(x) = x^2$

This function does not have one constant slope because its graph curves.

The derivative of this function is:

$f'(x) = 2x$

The notation:

$f'(x)$

is read as:

"f prime of x."

It represents the derivative of $f(x)$.

The original function tells us:

$f(x) = x^2$

The derivative tells us:

$f'(x) = 2x$

This means we now have a formula for finding the slope at different points on the curve.

At:

$x = 1$

$f'(1) = 2(1) = 2$

At:

$x = 2$

$f'(2) = 2(2) = 4$

At:

$x = 3$

$f'(3) = 2(3) = 6$

The slope becomes larger as x increases.

That matches what we would see on the graph: the curve becomes steeper.

The sign of a derivative also gives us useful information.

If:

$f'(x) > 0$

the function is increasing at that point.

If:

$f'(x) < 0$

the function is decreasing.

If:

$f'(x) = 0$

the graph is momentarily flat.

So derivatives help us answer more than:

"What is the slope?"

They also help us understand:

What is the function doing right now?

Differentiation Rules Made Understandable

Finding a derivative by repeatedly moving two points closer together would take a lot of time.

Fortunately, we have shortcuts called differentiation rules.

One of the most useful is the **power rule**.

If:

f(x) = x^n

then:

f'(x) = nx^(n - 1)

The rule is simple:

1. Bring the exponent down in front.
2. Subtract 1 from the exponent.

For example:

f(x) = x^3

Bring down the 3:

3x

Subtract 1 from the exponent:

3 - 1 = 2

So:

f'(x) = 3x^2

Another example:

$f(x) = x^5$

becomes:

$f'(x) = 5x^4$

Constants are even easier.

Suppose:

$f(x) = 10$

The number 10 never changes, no matter what x does.

So:

$f'(x) = 0$

A constant has a derivative of zero because its rate of change is zero.

We can also differentiate several terms separately.

Suppose:

$f(x) = x^3 + 2x^2 + 5x + 10$

Differentiate each term:

x^3 becomes $3x^2$

$2x^2$ becomes $4x$

5x becomes 5

10 becomes 0

Therefore:

$f'(x) = 3x^2 + 4x + 5$

More complicated functions require additional rules.

The **product rule** helps when two functions are multiplied.

The **quotient rule** helps when one function is divided by another.

The **chain rule** helps when one function is inside another.

For this book, the most important thing is not memorizing every rule immediately. It is recognizing why the rules exist.

Differentiation is simply the process of finding a derivative.

Once we have the derivative, we have a mathematical description of how the original function changes.

For example:

$f(x) = x^2$

tells us the original quantity.

While:

$f'(x) = 2x$

tells us its rate of change.

That distinction is the heart of derivatives.

Last Thoughts

A derivative answers one of the most important questions in calculus: how quickly is something changing right now? It takes the idea of slope and extends it from straight lines to curves, allowing us to measure change at individual points. Differentiation rules then give us faster ways to calculate those derivatives without rebuilding the calculation from limits every time. The important thing is to remember what the answer represents. A derivative is not simply another equation to memorize. It describes the behavior of the original function. In the next chapter, we will use derivatives to study motion, speed, acceleration, maximum and minimum values, and other practical problems.

Chapter 4: What Can We Actually Do With Derivatives?

Knowing how to find a derivative is useful, but the real value comes from knowing what we can do with it. Because derivatives measure rates of change, they can help us understand motion, identify when something reaches its highest or lowest point, and analyze how real quantities change over time. Instead of treating derivatives as equations we calculate and forget, we are going to use them to answer practical questions.

Understanding Motion, Speed, and Acceleration

Motion is one of the easiest ways to see derivatives in action.

Imagine a car traveling down a road. We could ask three different questions:

- Where is the car?
- How fast is the car moving?
- Is the car speeding up or slowing down?

Calculus treats these as related but different quantities.

Suppose the position of a car is:

$s(t) = t^2 + 2t$

Here:

s = position

t = time

If we want to know how quickly the car's position is changing, we find the derivative:

$s'(t) = 2t + 2$

This derivative represents **velocity**.

So we can write:

$v(t) = 2t + 2$

At:

$t = 2$

the velocity is:

$v(2) = 2(2) + 2$

$v(2) = 6$

The position function tells us where the car is. The velocity function tells us how quickly its position is changing.

But velocity can change too.

To find how quickly velocity is changing, we take another derivative:

$v'(t) = 2$

This gives us **acceleration**.

So we have:

Position -> Velocity -> Acceleration

Or mathematically:

$s(t)$ -> $s'(t)$ -> $s''(t)$

The notation:

$s''(t)$

is called the **second derivative** because we differentiated the original function twice.

This relationship is worth remembering:

- Position tells us where something is.
- Velocity tells us how quickly its position is changing.

- Acceleration tells us how quickly its velocity is changing.

Think about pressing the gas pedal in a car. Your speed may be 40 mph at one moment and 50 mph a few seconds later. Velocity tells us the speed and direction of movement, while acceleration describes how that velocity is changing.

This is a great example of an important calculus idea: **a rate of change can itself change**.

Finding Maximums and Minimums

Derivatives can also help us find the highest or lowest values of a function.

Suppose you throw a ball into the air.

Its height increases:

Higher.

Higher.

Higher.

Then, for a brief moment, it reaches the top.

After that, it falls.

At the very top, the ball changes from moving upward to moving downward. If we graph its height, the curve becomes flat at that highest point.

A flat slope means:

derivative = 0

This idea helps us find **maximums and minimums**.

A maximum is a high point.

A minimum is a low point.

Consider:

$f(x) = -x^2 + 6x + 5$

Differentiate:

$f'(x) = -2x + 6$

To look for a possible maximum or minimum, set the derivative equal to zero:

$-2x + 6 = 0$

Solve:

$-2x = -6$

$x = 3$

Now put 3 into the original function:

$f(3) = -(3^2) + 6(3) + 5$

$f(3) = -9 + 18 + 5$

$f(3) = 14$

So the important point is:

(3,14)

Why is this a maximum?

Before $x = 3$, the function is increasing.

After $x = 3$, the function is decreasing.

Think of climbing a hill:

Increasing -> Top -> Decreasing

The top is the maximum.

For a minimum, the pattern would be:

Decreasing -> Bottom -> Increasing

Finding the highest or lowest possible value in a situation is called **optimization**.

Optimization problems might ask questions such as:

- What production level produces the greatest profit?
- What dimensions use the least material?

- When does an object reach its greatest height?
- What value produces the lowest cost?

The details change, but derivatives give us a way to locate the important turning points.

One caution is important: derivative = 0 gives us a point to investigate. It does not automatically mean we have found a maximum or minimum. We still need to understand what the function is doing around that point.

Using Rates of Change to Understand Real Problems

Derivatives are useful because many real situations involve quantities that change.

Consider a business with total cost:

C(q) = 500 + 10q + 0.05q^2

Here:

C = total cost

q = quantity produced

Differentiate:

C'(q) = 10 + 0.1q

Suppose the company is producing:

$q = 100$

Then:

$C'(100) = 10 + 0.1(100)$

$C'(100) = 20$

What does 20 mean?

It does not mean the company's total cost is $20.

The original function gives us total cost.

The derivative tells us **how quickly total cost is changing** around that production level.

At 100 units, the model says cost is increasing at approximately $20 for an additional unit of production.

Now consider temperature.

Suppose:

T(t) = 80 - 4t + 0.2t^2

Differentiate:

$T'(t) = -4 + 0.4t$

At:

$t = 2$

$T'(2) = -4 + 0.8$

$T'(2) = -3.2$

The negative answer tells us something important.

The temperature is decreasing at that moment.

This gives us a simple way to interpret derivatives:

- Positive derivative: the quantity is increasing.
- Negative derivative: the quantity is decreasing.
- Zero derivative: the quantity is momentarily not changing.

The same thinking can be applied to many situations.

If $P(t)$ represents population, $P'(t)$ describes population change.

If $R(q)$ represents revenue, $R'(q)$ describes how revenue changes as quantity changes.

If $T(t)$ represents temperature, $T'(t)$ describes how temperature changes over time.

If $s(t)$ represents position, $s'(t)$ describes velocity.

Whenever you calculate a derivative, ask:

"What is changing, and what does this number tell me about that change?"

That question is often more important than the calculation itself.

Last Thoughts

Derivatives become much more useful when we stop viewing them as formulas and start using them to understand behavior. They can connect position to velocity and velocity to acceleration, help us identify maximum and minimum values, and describe how quantities such as cost, temperature, revenue, or population are changing. The key is always to interpret the derivative in the context of the problem. So far, calculus has mostly taken us from a quantity to its rate of change. In the next chapter, we will reverse that thinking and explore integrals, which help us understand what happens when many small amounts of change are added together.

Chapter 5: Integrals: Adding Up Tiny Pieces

So far, calculus has helped us move from a quantity to its rate of change. If we know position, a derivative can give us velocity. If we know a cost function, a derivative can tell us how quickly cost is changing. Integrals allow us to look at the problem from the other direction. Instead of asking how quickly something is changing, we can ask how much has accumulated. An **integral** combines many small pieces to help us find a larger total.

Understanding Accumulation One Small Piece at a Time

Imagine water flowing into a tank at a constant rate of:

5 gallons per minute

After 4 minutes, the amount added is:

5 x 4 = 20 gallons

That is easy because the rate never changes.

But what if the water flow changes every moment?

Maybe it starts at 5 gallons per minute, increases to 6, then 7, then falls back to 5.

Now there is no single rate we can multiply by the total time.

One solution is to divide the time into small pieces.

For each piece, we estimate:

amount added = rate x small amount of time

Then we add all those amounts together.

We could use:

- 1-minute pieces
- 30-second pieces
- 1-second pieces
- Even smaller pieces

The smaller the pieces become, the more closely our calculation can follow the changing rate.

That is the basic idea behind integration:

Break something into tiny pieces, then add those pieces together.

The integral symbol looks like this:

$\int$

You can think of it as representing a continuous sum.

Suppose:

r(t)

represents the rate at which water enters a tank.

Then an expression such as:

∫from 0 to 5 r(t) dt

asks us to accumulate that rate from time 0 to time 5.

In plain English:

"How much water entered during this period?"

This same idea works in many situations.

If we know a changing velocity, we can accumulate movement.

If we know a changing flow rate, we can accumulate water.

If we know a changing rate of energy use, we can accumulate energy consumption.

The situation changes, but the question stays similar:

How much do all these little pieces add up to?

Area Under a Curve and the Definite Integral

Integrals also have a useful visual meaning.

Suppose:

f(x) = 3

from:

x = 0

to:

x = 4

If we graph this function, the area underneath it forms a rectangle.

The rectangle has:

height = 3

width = 4

So:

Area = 3 x 4 = 12

We can write the same calculation as:

∫from 0 to 4 3 dx = 12

Because we have specified a starting point and an ending point, this is called a **definite integral**.

Now suppose:

f(x) = x^2

The graph is curved.

We cannot simply use:

length x width

because the height keeps changing.

Instead, imagine filling the area beneath the curve with narrow rectangles.

With only a few rectangles, our estimate may be rough.

Add more rectangles and make them narrower, and they fit the curve better.

Make them narrower still, and the approximation improves even more.

Integration allows us to consider what happens as those pieces become extremely small.

Once again, limits are working in the background.

A definite integral therefore gives us a way to calculate accumulation across an interval.

There is one important detail.

If a graph is above the x-axis, its contribution to the integral is positive.

If it falls below the x-axis, its contribution is negative.

This matters when the function represents something with direction, such as velocity.

Suppose you walk:

10 feet forward

and then:

4 feet backward

Your total distance traveled is:

$10 + 4 = 14$ feet

But your final change in position is:

$10 - 4 = 6$ feet

An integral of velocity naturally accounts for those positive and negative directions.

So a definite integral does not always mean ordinary geometric area. More broadly, it represents **accumulation over an interval**.

Antiderivatives and Indefinite Integrals

There is a much easier way to calculate many integrals than adding thousands of tiny pieces by hand.

Remember differentiation.

Suppose:

$f(x) = x^3$

Its derivative is:

$f'(x) = 3x^2$

Now reverse the question.

If we start with:

$3x^2$

what function did it come from?

One answer is:

x^3

This reverse process gives us an **antiderivative**.

Differentiation moves:

$x^3 \to 3x^2$

Antidifferentiation moves:

$3x^2 \to x^3$

For powers of x, we can use a simple rule.

To integrate:

x^n

we:

1. Add 1 to the exponent.
2. Divide by the new exponent.

For example:

$\int x^2\,dx$

Add 1:

$2 + 1 = 3$

Then divide by 3:

$\int x^2\,dx = x^3/3 + C$

The:

$+C$

is important.

Suppose:

$f(x) = x^3$

Its derivative is:

$3x^2$

But:

$f(x) = x^3 + 10$

also has the derivative:

$3x^2$

So does:

$f(x) = x^3 - 50$

Differentiation makes constants disappear.

That means when we work backward, we do not know what constant might have been there.

The **constant of integration**, written as C, represents that unknown constant.

An integral without specified starting and ending values is called an **indefinite integral**.

For example:

$\int 2x \, dx = x^2 + C$

We can check the answer by differentiating:

$d/dx(x^2 + C) = 2x$

Now consider a definite integral:

$\int$from 0 to 3 2x dx

First find an antiderivative:

x^2

Then evaluate it at 3:

$3^2 = 9$

Evaluate it at 0:

$0^2 = 0$

Subtract:

$9 - 0 = 9$

Therefore:

$\int$from 0 to 3 2x dx = 9

This gives us a much faster way to calculate accumulated quantities.

It also reveals something important.

Derivatives and integrals appear to work in opposite directions.

A derivative can take a quantity and reveal its rate of change.

An integral can take a rate and help us recover accumulated change.

That connection is not an accident. It is one of the biggest ideas in all of calculus.

Last Thoughts

Integrals help us answer a simple but powerful question: how much has accumulated? We can imagine integration as breaking a quantity into tiny pieces and adding those pieces together. Definite integrals measure accumulation across a particular interval, while indefinite integrals help us find antiderivatives. Most importantly, integration begins to reverse what differentiation does. Derivatives move us toward rates of change, while integrals can move us back toward accumulated quantities. In the next chapter, we will put these ideas together and see why derivatives and integrals are so closely connected.

Chapter 6: Putting Calculus Together

We have now covered the main building blocks of calculus. Limits help us understand what happens as values get closer to something. Derivatives measure rates of change. Integrals measure accumulation. The final step is understanding how these ideas fit together. In particular, derivatives and integrals have a powerful relationship: one can often undo what the other does. Once you understand that connection, calculus starts to feel less like a collection of separate rules and more like one system for understanding change.

How Derivatives and Integrals Are Connected

Think about driving a car.

Suppose:

$s(t)$

represents your position over time.

If we take its derivative:

$s'(t)$

we get velocity.

So:

Position -> Derivative -> Velocity

The derivative tells us how quickly position is changing.

Now suppose we start with velocity instead.

If we know how fast you are moving throughout a trip, we can use an integral to accumulate that movement and determine your change in position.

So we can move the other way:

Velocity -> Integral -> Change in Position

Here is a simple example.

Suppose:

$s(t) = t^2$

Differentiate:

$s'(t) = 2t$

So our velocity is:

$v(t) = 2t$

Now suppose we want to know how much position changes between:

$t = 1$

and:

$t = 3$

Using the position function:

$s(3) = 9$

$s(1) = 1$

Therefore:

change in position = 9 - 1 = 8

Now let's use the velocity instead:

∫from 1 to 3 2t dt

An antiderivative of 2t is:

t^2

Evaluate it at 3:

3^2 = 9

Evaluate it at 1:

1^2 = 1

Subtract:

9 - 1 = 8

We got the same answer.

That shows us the connection:

- Differentiation can take us from a quantity to its rate of change.
- Integration can take a rate of change and recover accumulated change.

This relationship works far beyond motion.

If:

$C(q)$

represents total cost, then:

$C'(q)$

describes how quickly cost changes as production changes.

Integrating $C'(q)$ across a range of production levels can tell us how much total cost changed across that range.

Different situation, same calculus.

Understanding the Fundamental Theorem of Calculus

The connection we just discovered has an important name: the **Fundamental Theorem of Calculus**.

Despite the intimidating name, its main idea is straightforward:

Differentiation and integration are closely connected reverse processes.

Suppose:

$F'(x) = f(x)$

This tells us that f(x) is the derivative of F(x).

Then:

∫from a to b f(x) dx = F(b) - F(a)

In simpler terms:

1. Find an antiderivative.
2. Evaluate it at the ending point.
3. Evaluate it at the starting point.
4. Subtract.

Let's try it.

Suppose:

f(x) = 3x^2

We want:

$\int_{1}^{2} 3x^2 \, dx$

An antiderivative is:

$F(x) = x^3$

Now evaluate the ending point:

$F(2) = 2^3 = 8$

Evaluate the starting point:

$F(1) = 1^3 = 1$

Subtract:

$8 - 1 = 7$

Therefore:

$\int_{1}^{2} 3x^2 \, dx = 7$

Without this connection, we would have to think about adding increasingly tiny pieces underneath the curve.

The Fundamental Theorem gives us a much easier method.

It also explains why we spent time learning both derivatives and integrals.

They are not two unrelated topics.

Suppose A(t) represents the total amount of water that has accumulated in a tank.

Then:

A'(t)

tells us how quickly water is entering or leaving at a particular moment.

If we instead know the rate at which water is entering, integrating that rate over time tells us how much the amount of water changed.

So:

Accumulated amount -> Derivative -> Rate

and:

Rate -> Integral -> Accumulated change

That relationship sits at the heart of introductory calculus.

Seeing Calculus in Motion, Growth, Science, Economics, and Everyday Life

Now that we understand the major ideas, we can see why calculus appears in so many different subjects.

Motion

For motion:

Position -> Velocity -> Acceleration

Each arrow can represent differentiation.

If we know acceleration and want to work backward toward changes in velocity, integration becomes useful.

Growth

Suppose:

P(t)

represents a population.

Then:

P'(t)

tells us how quickly the population is changing.

A positive derivative means the population is increasing.

A negative derivative means it is decreasing.

Integration can help determine how much population change accumulates over a period.

Science

Scientists constantly study changing quantities.

Calculus can help describe:

- Changing temperature
- Movement and acceleration
- Population growth
- Changing concentrations
- Flow rates
- Energy and other accumulated quantities

The specific equations may become much more advanced, but the underlying questions are familiar.

How quickly is something changing?

How much change has accumulated?

Economics

Calculus also helps economists and businesses analyze changing quantities.

Suppose:

$C(q)$ = total cost

Then:

$C'(q)$

tells us how cost changes as production changes.

Suppose:

$R(q)$ = revenue

Then:

$R'(q)$

tells us how revenue changes as quantity changes.

Derivatives can also help locate maximum or minimum values, which makes them useful for optimization problems.

Everyday Life

You encounter the ideas behind calculus even when you are not solving equations.

A speedometer measures a rate of change.

Rainfall can accumulate over time.

Electricity use accumulates throughout the month.

A vehicle speeds up and slows down.

Money can grow at changing rates.

A hot drink cools over time.

Calculus gives us a mathematical language for studying these kinds of changes.

When you encounter a calculus problem, it can help to ask a few simple questions:

- Is something approaching a value? Think about limits.
- Am I looking for how quickly something changes? Think about derivatives.
- Am I looking for a maximum or minimum? Derivatives may help.
- Am I adding up changing amounts? Think about integrals.
- Am I trying to recover accumulated change from a rate? Think about integrals.

You may eventually encounter much more complicated calculus, but these questions provide a foundation for understanding what the mathematics is trying to accomplish.

Last Thoughts

Calculus becomes easier to understand when you see the connections instead of treating every formula as a separate rule. Limits help us understand approaching behavior, derivatives describe rates of change, and integrals describe accumulation. The Fundamental Theorem of Calculus connects derivatives and integrals, showing us that the two major operations are deeply related. Whether calculus is being used to study motion, growth, science, economics, or an everyday changing quantity, the basic questions remain remarkably similar: **What is changing? How quickly is it changing? And how much change has accumulated?** Once you can recognize those questions, you have understood the central idea behind calculus.

Last Thoughts

Calculus can look complicated because of its symbols and formulas, but the ideas underneath them are much easier to understand. Throughout this book, we have reduced calculus to a few central questions about **change, rates, and accumulation**. You do not need to look at every equation and immediately know what to do. What matters first is recognizing what kind of problem the mathematics is trying to solve.

Calculus Is a Way of Thinking About Change

Calculus begins with change.

A car changes position.

A population grows.

A business's costs change as production increases.

Water accumulates in a container.

A temperature rises or falls.

Once something changes, we can begin asking mathematical questions about that change.

Limits ask:

"What value are we approaching?"

Derivatives ask:

"How quickly is this changing?"

Integrals ask:

"How much has accumulated?"

Those three questions give you a useful way to organize much of what you learned in this book.

You can also see how the ideas connect:

- **Limits** help us examine behavior as values get closer and closer.
- Derivatives use this idea to measure change at a particular point.
- Integrals combine small pieces to measure accumulation.
- The Fundamental Theorem of Calculus connects differentiation and integration.

Instead of thinking of calculus as hundreds of disconnected formulas, think of it as a collection of tools for studying changing quantities.

Understanding Before Memorizing

Formulas are important in calculus, but memorizing a formula does not automatically mean you understand what it does.

Consider:

$f'(x)$

You now know that this represents a derivative.

But more importantly, you know to ask:

"What is changing?"

If you see:

$s'(t)$

and s represents position, you can recognize that the derivative describes how position changes with time.

That leads us toward velocity.

If you see an integral such as:

∫from a to b f(x) dx

you know that we are accumulating something across an interval.

That understanding gives the notation meaning.

When working through calculus problems, try using this order:

1. Identify the quantities involved.
2. Determine what is changing.
3. Decide what the problem is asking you to find.
4. Choose the calculus tool that matches the question.
5. Perform the calculation.
6. Interpret what your answer actually means.

That final step is especially important.

An answer of:

5

means very little by itself.

Does it mean 5 miles?

5 miles per hour?

5 dollars per additional unit?

5 gallons accumulated?

The mathematics produces a number, but understanding the problem tells you what that number represents.

That is why understanding should come before memorization.

Where to Go After You Understand the Fundamentals

This book introduced the foundation, but calculus extends much further.

Once you are comfortable with the ideas covered here, you can continue into topics such as:

- More advanced differentiation techniques
- More advanced integration techniques
- Trigonometric and exponential functions
- Related rates
- More complex optimization problems
- Sequences and series
- Multivariable calculus
- Differential equations

You do not need to tackle all of these at once.

A good next step is simply to practice the fundamentals until you can look at a problem and recognize what it is asking.

When you see a changing quantity, ask:

"Do I need its rate of change?"

If yes, you are probably thinking about a derivative.

Ask:

"Do I need to add up changes or recover an accumulated quantity?"

If yes, you are probably thinking about an integral.

Ask:

"Do I need to understand what happens as something gets extremely close to a value?"

If yes, you are probably thinking about a limit.

That ability to recognize the problem is a major part of learning calculus.

You may forget a differentiation rule.

You may need to look up an integration formula.

You may need additional practice solving equations.

That is fine.

The formulas can be reviewed.

The more important foundation is knowing **why you are using them**.

Last Thoughts

Calculus does not have to begin with complicated notation. It can begin with a simple observation: things change, and mathematics gives us ways to understand that change. You now have a foundation for understanding limits, derivatives, rates of change, optimization, integrals, and the connection between differentiation and integration. There is much more calculus to learn if you choose to continue, but you no longer have to approach the subject as a collection of mysterious symbols. Start with the question, understand what is changing, and then use the mathematics to describe it.

About the Author

Roggie "Raj" Clark is an entrepreneur and a 10-year marketing professional with a proven track record in e-commerce and digital business growth. Having built and managed successful online stores with thousands of sales and consulted for numerous marketing clients, Roggie has mastered the intricacies of online marketing, branding, and consumer behavior. He has also written several popular books on digital marketing, SEO (search engine optimization), and programming (HTML, CSS, and JavaScript).

As a former marketing manager at a fintech company—a role he secured through his own marketing and financial expertise—Roggie gained firsthand experience in financial strategy and wealth-building. This journey inspired him to share insights with amateur investors, helping them save money, make smarter financial decisions, and build long-term wealth. Raj currently lives in Maryland with his sister and niece.

www.ingramcontent.com/pod-product-compliance
Lightning Source LLC
LaVergne TN
LVHW050827061126
851857LV00012B/1015
* 9 7 9 8 2 5 6 2 9 2 4 1 6 *